Ecosystems & Environment

ANN FULLICK

Heinemann Library
Des Plaines, Illinois

Customer Service 888-454-2279

Designed by AMR
Illustrations by Art Construction
Printed in Hong Kong

04 03 02 01 00
10 9 8 7 6 5 4 3 2

Library of Congress Cataloging-in-Publication Data

Fullick, Ann, 1956-
    Ecosystems & environment / Ann Fullick.
        p. cm. – (Science topics)
    Includes bibliographical references and index.
    Summary: Discusses ecosystems and the environment, including
habitats, food chains and food webs, adaptation, human impact, and
genetic engineering.
    ISBN 1-57572-775-7 (library binding)
    1. Ecology Juvenile literature. [1. Ecology.] I. Title.
    II. Title: Ecosystems and environment. III. Series.
    QH541.14.F85 1999
    577—dc21   99-12867
                                    CIP

Acknowledgments
The Publishers would like to thank the following for permission to reproduce photographs:
FLPA Images of Nature/K. Aitken, p. 4; Science Photo Library/CNRI, p. 5; FLPA Images of Nature/
F. Hoogervorst/Foto Natura, p. 6; NHPA/Hellio & Van Ingen, p. 7; FLPA Images of Nature/Fritz
Folking, pp. 8, 9; Holt Studios/Nigel Cattlin, pp. 11, 24; NHPA/David Middleton, p. 12; Science
Photo Library/Eye of Science, p. 13; Ancient Art & Architecture, p. 16; Science Photo Library/
Claude Nuridany & Marie Perennou, p. 17; Holt Studios/Mary Cherry, p. 18; Topham/Associated
Press, p. 19; NHPA/Nigel J. Dennis, p. 20; Tony Stone Images/Ken Fisher, p. 22; Quadrant Picture
Library, p. 23; Morris, Frances & Richard, p. 25; PA News Photo Library/Roslin Institute, p. 27;
Mary Evans Picture Library, p. 29.

Cover photograph reproduced with permission of NHPA/B. Jones and M. Shimlock.

Our thanks to Geoff Pettengell for his comments in the preparation of this book.

Every effort has been made to contact copyright holders of any material reproduced in this
book. Any omissions will be rectified in subsequent printings if notice is given to the Publisher.

Any words appearing in the text in bold, **like this**, are explained in the Glossary.

Contents

Habitats and Ecosystems

Every living thing has a home, and all living things interact with their **environment** (surroundings) and with each other. By finding out about these relationships, we can understand more about the way our planet works.

Home sweet home

Almost everywhere in the world is home to at least one type of living organism. Some of these **habitats** are large and easy to think of—a coral reef, a rotting log, and a meadow by a stream are examples. Other habitats are more unusual and far less obvious. For example, the clown fish lives among the poisonous stinging tentacles of large sea anemones, while another fish, *Tilapia grahami*, survives in the warm mineral waters of Lake Magadi in Kenya at temperatures of up to 109°F (42.8°C).

In an area of the sea like this, there are a number of different habitats that provide homes for a wide range of living organisms.

A wider view

If we look at a single plant or animal in its habitat, we soon discover that its home is only a small part of a particular area. Within that area, there are a lot of other species of animals and plants, and each lives in its own habitat. Very often, these animals and plants interact with each other. For example, in a forest, the trees are the habitat of many different species of birds, mammals, and insects. Fallen trees are the habitat of other creatures like wood lice, **bacteria**, and fungi.

The forest floor is also home to many plants and animals. Some of the animals eat parts of the plants, while others eat animals. The kinds of plant that can grow in a woodland depend on the amount of rainfall in the area, the type of rocks and soil, the amount of light, and the normal temperature range. Factors like these also affect the types of animals found, both directly and indirectly, as a result of the plants that are able to grow.

his complex interaction of many
ifferent species of living organisms with
ne **abiotic** (nonliving) features of their
ome is known as an **ecosystem**. The
udy of ecosystems is called **ecology**.
biotic factors are very important in an
cosystem. They not only determine the
ind of organisms living in an area, they
lso affect the numbers of each type.

The human habitat

surprising number of organisms prefer
live on human beings rather than on any
ther mammal. Every healthy human being
home to countless bacteria. They live all
ver the surface of our skin and in our
igestive system. Mites and other tiny
nvertebrates swing through our body
hair like the inhabitants of a microscopic
jungle. We are also hosts to parasitic
worms, fungi, lice, fleas, and many other
organisms that wait for a breakdown in
our body's defenses. As soon as this
happens, they can make themselves at
home on or in our warm, food-rich bodies.

Some of the many millions of habitats available
in the world are on our own bodies. Organisms
like this head louse are quick to take advantage
of the warmth and ready food supply.

Food Chains

All living things need food to give them energy. For most living organisms, the energy from food originally came from the sun, either directly or indirectly.

Powerhouse of the planet

Plants are often taken for granted, but they are the true powerhouse of our planet. Living things need energy to live, grow, and reproduce. Plants and some **bacteria** are the only living organisms that can harness the sun's energy and use it to provide themselves with food.

Plants do this using a process called photosynthesis. Energy from the sun captured using a special green coloring called **chlorophyll**, which is found in **chloroplasts** in the cells of leaves. Once captured, the energy is used to join carbon dioxide from the air with water from the soil to make sugar. Oxygen is released as a waste product. The sugar is then used to provide energy for the plant and raw material for building new plant cells. Because this ability to build new biological material (**biomass**), plants are known as producers.

All of the biomass present in a tree comes from photosynthesis. Food and oxygen are produced in this one amazing process. Plants are often taken for granted, yet they are vital for life on Earth.

Links in a chain

Because of their unique ability to produce new biomass, plants are used as food, either directly or indirectly, by all of the different types of animals on Earth. Animals are called **consumers** because they have to eat other organisms to live. They either eat plants themselves or eat animals that are linked back to other animals that eat plants. These feeding relationships, which always lead back to plants, are known as food chains.

Herbivores eat only plants and are known as **primary consumers**. Animals that eat herbivores are called **secondary consumers**. Secondary consumers are themselves eaten by **tertiary consumers**. Animals that hunt and eat other animals are known as **predators**, while the hunted—mainly herbivores and small **carnivores**—are known as **prey**.

Food chains can be very short. For example, grass is eaten by cows and cows are eaten by people. They can

A carnivore may not seem to have much to do with plants, but the moorhen doomed to be a meal for this fox was busy feeding on plants only a short time ago. Without those plants, there would be no food for the meat eaters.

also be quite long and move through many organisms of increasing size. Whatever the length of a food chain, it will always start with a plant. All food chains also include **decomposers**. These are specialized consumers, often fungi or bacteria, which break down the tissues of a plant or animal when it dies and return the nutrients to the soil.

Freedom from the sun

A very tiny group of living organisms do not rely on green plants and the sun to supply their energy. Specialized bacteria known as methanobacteria produce energy from hydrogen and carbon dioxide. Methane gas is also produced, but as a waste product. Other bacteria use sulfur or hydrogen sulfide, ammonium ions, and nitrites as energy sources. The strange biochemistry of these organisms allows them to live in difficult habitats such as swamps with extremes of temperature, salt concentration, and pressure. However, it has not allowed them to take the place of plants in providing energy for all other living organisms on the planet.

Food Webs

A **food chain** gives us a simple model of the feeding relationships between plants and animals. However in real life, the situation is almost always more complicated.

Living webs

Within an **ecosystem**, there are many different types of plants and animals living in a particular area. When we construct a food web, we identify the feeding relationships between the **producers** and **consumers** in an ecosystem and show how the animals and plants are interlinked.

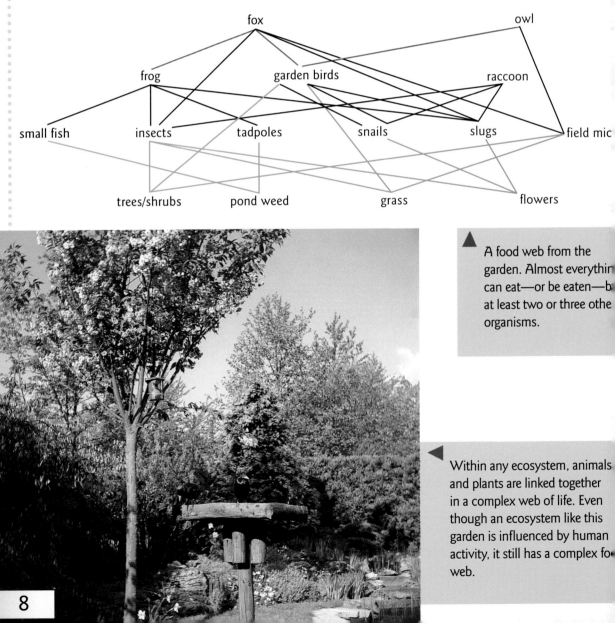

A food web from the garden. Almost everythin[g] can eat—or be eaten—b[y] at least two or three othe[r] organisms.

Within any ecosystem, animals and plants are linked together in a complex web of life. Even though an ecosystem like this garden is influenced by human activity, it still has a complex fo[od] web.

Natural buffering

If each animal had only a single food source, then the whole balance of nature would be very delicate indeed. For example, if a **herbivore** ate only one type of plant or a **carnivore** hunted only one type of **prey**, they would find it difficult to survive.

However, as food webs show, this is rarely the case. In theory, if all of the slugs in a garden were removed by a clever gardener, then more plants would survive. In fact, there would just be more food left for the greenflies, ants, and other organisms that eat plants. The many garden birds, raccoons, and other animals that eat slugs would not starve to death. Instead of the slugs, they would eat more of the other species they like. As a result, there would probably be fewer worms, beetles, and other creepy-crawlies in the garden.

An environmental event that affects one organism does not have as large an effect on a whole ecosystem as we might expect. This is because of the buffering effect of the many organisms that are all interacting.

Panda problems

Pandas are a well-known symbol of conservation because they are themselves dangerously close to **extinction**. Part of the problem for pandas is that they are not really part of a food web. Instead, they are part of a simple food chain in which pandas eat bamboo and little else. When the bamboo plants all flower and die, which happens every hundred years or so, and if the pandas' habitat is destroyed, they immediately face starvation. By looking at how vulnerable the panda is, we can certainly appreciate the buffering effect of food webs. Fortunately, they support the majority of living organisms.

Pandas are beautiful but vulnerable. These rare animals have the digestive system and teeth of a carnivore but eat a diet of almost entirely one plant. No one is sure why pandas eat only bamboo, but it certainly makes it much harder for them to survive.

Pyramids of Life

A close look at the organisms that make up a **food chain** or **food web** reveals an important principle of **ecology**: in almost every community, there are more plants than **herbivores** and more herbivores than **carnivores**.

SCIENCE ESSENTIALS

Pyramids of numbers show the numbers of organisms at each level of a food chain.
Pyramids of biomass show the amount of biological material in the organisms at each level of a food chain.

Pyramids of numbers

Animals and plants that live in any particular **ecosystem** are often linked by food chains or food webs.
For example:

> grass → rabbits → fox
> sea plants → fish → seals → polar bears

Although food chains like these show us the feeding relationships between the organisms, they do not give us any idea of how many of the different organisms there are. If we count or quantify the number of organisms in a food chain, or in the chains that make up a complex web, we can build up a pyramid of numbers.

The numbers in many food chains confirm what we expect: many plants at the bottom of the pyramid support fewer plant eaters, which in turn feed a smaller number of larger **secondary consumers**. These may, in turn, feed an even smaller number of top **predators**.

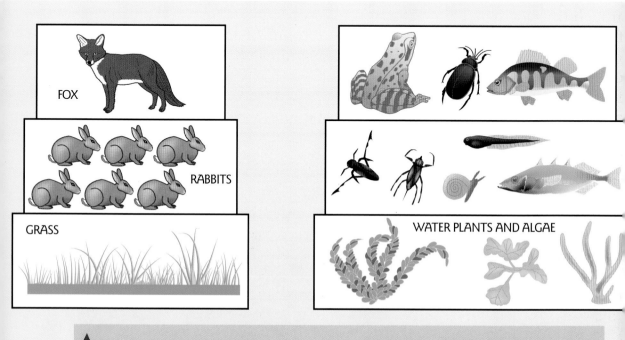

FOX

RABBITS

GRASS

WATER PLANTS AND ALGAE

▲ Whether we look at a single food chain or a whole food web, the numbers of living organisms supported at each level of a pyramid of numbers tend to get smaller.

...omass is better

...metimes there is a problem when ...e use pyramids of numbers to help ...alyze an ecosystem. The numbers ...n be greater at the top, so there is ... pyramid.

Thousands of aphids feed on just one rosebush. The aphids themselves act as food for ladybugs and other predators. The pyramid of numbers for a food chain like this is a very strange shape indeed!

...owever, when we look at the ...mount of actual biological material ...**iomass**) at each level of a food ...nain, we always get the picture we ...xpect. We see the amount of biomass ...lling as we move from plants to ...erbivores to carnivores. Why do the ...vels of biomass fall in this way? ...Where does all the "lost" material go?

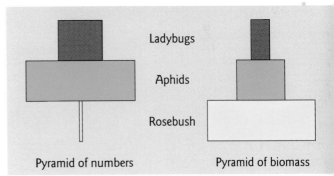

Ladybugs

Aphids

Rosebush

Pyramid of numbers   Pyramid of biomass

...luch of the sun's energy captured by ...lants in **photosynthesis** does not get ...nrough the food chain. A great deal ...f the plant material eaten by ...erbivores is passed out of their ...odies as **feces** because they cannot ...igest **cellulose**. Carnivores also leave ...arts of their **prey** uneaten and get rid

of undigested material as feces. The food that animals do digest is used to provide them with energy for movement and, if the animal is a mammal, to produce body heat. Only a small amount of the food energy is turned into new animal, to be passed on to the next animal in the chain.

Measuring biomass

...ounting numbers of living organisms ... a food chain can be difficult, but ...easuring biomass is even harder. ...the animals and plants are alive, their ...omass still contains water. A measure ...f their wet biomass can be very ...accurate. For example, it will be

affected by how much water the animals have drunk. Measuring dry biomass gives an accurate picture of a food chain. Unfortunately, to get a dry biomass, the organisms have to be killed and dried. This destroys the chain being studied!

Making the Best of It

We can find a whole range of different **habitats** on any small area of the earth's surface. Wherever life is possible, it will be found. Animals and plants have **evolved** over millions of years to take advantage of any opportunities that come their way.

Survival!

Animals and plants need to be able to survive and reproduce in the **environment** in which they live.

Many have developed adaptations that enable them to take advantage o the food sources available.

Desert organisms have many differen ways of surviving. Cacti have long, deep roots and stems full of cells that can store water. The leaves of cacti are thin spines, which reduce the water loss. A special form of **photosynthesi** means that they only need to take in carbon dioxide during the night and keep their pores closed during the day to prevent water loss. All of these adaptations mean that cacti do well i conditions where most plants would wilt and die.

Other desert plants survive by avoiding most of the dry conditions and living as **dormant** seeds that onl start to grow after rain has fallen. Once wet, they flower and set new seeds within a few days. The plants will survive even if there is no more rain for several years.

The hot, dry conditions of a desert environment a too challenging for most living things.
The organisms that can survive and reproduce in these extreme conditions have special adaptations that make their survival possible.

the animal world, camels are well [k]own for their humps of fatty tissue. [Th]ese provide food for the camels for [m]any days. Camels are able to [m]anage for long periods without [wa]ter. They have a special system for [co]oling the blood as it flows to the [br]ain and **enzymes** that tolerate far [hi]gher temperatures than those of [m]ost mammals. Their thick insulating [co]ats trap a layer of air to keep out the [he]at of the day and the cold at night.

Their ears can close to prevent sand from getting in and their special wide feet are excellent for walking on sand. Camels are perfectly adapted to their environment. Other desert animals have different adaptations. Many of them simply avoid the heat of the day by living in underground burrows and only emerging during the evening, night, and early morning. Most desert animals are adapted to survive with little water.

[H]abitats everywhere

[Li]ving things have made their homes [in] many different environments— [fr]om the cold regions of the Arctic [an]d Antarctic to the arid deserts of the [w]orld, from the heights of the Rocky [M]ountains to the depths of the [M]ariana Trench (the deepest known [pa]rt of the ocean). Some of their [ad]aptations are fairly common such as [fu]r, which gives them an insulating [la]yer, or different shaped mouths, [be]aks, and teeth that allow them to [ca]tch and eat different types of food. [Ot]her adaptations are quite bizarre.

[Th]e mole rat of East Africa is an [un]usual creature. It spends its entire [lif]e underground and only eats the [ro]ots of grass. Family units live [to]gether in a maze of underground

tunnels with special chambers for different activities. Their adaptations to this life are dramatic. They are blind, naked, sausage-shaped animals with grey, wrinkled skin. They have huge incisor teeth, which they use to move soil while their mouths are shut. By eating roots and living completely underground, they avoid dangerous **predators**, unlike the prairie dogs and rabbits that feed on grass at the surface.

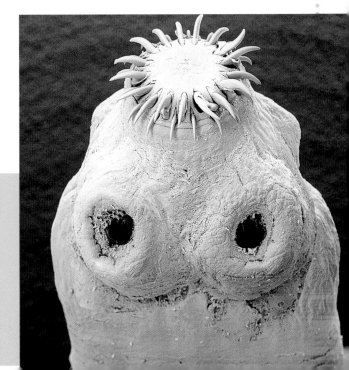

▶ The ultimate adaptation? Parasitic organisms like this tapeworm live inside the body of another organism. Warm and protected from predators, they feed on already-digested food. To avoid being digested themselves, they have to be adapted for hanging tightly onto their host. These parasites have certainly found a very exclusive habitat.

Keeping up with Change

Organisms must be adapted for the **habitat** in which they live. Sometimes this involves adapting to cope with change. Changes in the **environment** can test an organism to the limits. Those organisms that cannot cope often die as a result. The adaptable ones survive and breed.

All things change!

Some organisms live in very stable environments, but others have to cope with constant change. One of the best examples of organisms coping with change is in the tidal zones of a coastline. Within 24 hours, the tide comes in and goes out at least twice. As a result, the plants and animals are covered by moving water for part of the time. Next, they are exposed either on the sand or in the shallow water of a rock pool. As the water in a rock pool evaporate in the heat of the sun, its salinity (concentration of salt) becomes far greater than that of normal seawater.

▼
Habitats such as rock pools are a real challenge the organisms that live in them. A rock pool forms a **microhabitat**. Depending on its positic on the beach, the microhabitat can be out of th moving sea from a few minutes to a few hours every day. The trapped plants and animals have to cope with the changes in temperature and salinity.

Adaptations to cope with the hardships of rock pool life vary. Some organisms, particularly shellfish, simply trap some seawater in their shells and clamp down to wait until the tide returns. Others, like crabs and seaweed, have **evolved** tissues that can cope with the increase in temperature and salinity that occurs when a rock pool is exposed to the sun for some time. Any organisms trapped in a rock pool that are not adapted to cope will probably die before the tide returns.

The changing seasons

Near the equator, the weather stays very similar all through the year. But in many other parts of the world, living organisms have to cope with dramatic changes in conditions. Within one year in some states, plants and animals have to adapt to changes of about twelve hours in the amount of daylight and to temperatures that range from 14°F (-10°C) in winter to up to 86°F (30°C) in summer. The adaptations that enable organisms to cope range from physical characteristics to big changes in behavior.

Hibernation: Some animals build up large fat stores to get ready for the cold weather. During the winter, they drop their body metabolism to a very low level and sleep. Others indulge in partial hibernation and wake every so often to feed on stores of food.

Dormancy: Many plants lose their leaves and shut down for the winter. Others exist as a dormant seed, bulb, or tuber underground.

Migration: Many birds and some mammals travel vast distances to avoid adverse conditions and migrate back again when the seasons change.

Coat changes: Some animals grow a thin coat for the warm part of the year and a much thicker one for the winter. Others completely change the color of their feathers or fur to give themselves maximum camouflage throughout the year.

Biological clocks

Almost all living organisms have to cope with the 24-hour rhythm of day and night. Many organisms have their own internal rhythms that are linked to this 24-hour cycle. These are known as biological clocks, and they almost all keep a **circadian rhythm**. Our human clocks tell us when to sleep. They also control the levels of many chemicals in our bodies.

Agents for Change

Most of the **adaptations** we see in animals and plants have **evolved** in response to changes in the natural world around them. Changes in the **environment** are not always the result of natural phenomena. There is a new agent for change—the human race.

The human factor

Throughout the long history of life on Earth, species of animals and plants have been in a constant process of adapting and changing to survive. Species that are not well adapted become extinct. When a new **habitat** becomes available, animals and plants will adapt to fill it.

Usually, environmental changes occur slowly, over a long period of time. This gives organisms a chance to adapt over many generations.

In the relatively short time since human beings have evolved, we have had a major impact on the environment and the animals and plants that share our planet. We have reached a point where we can bring about major environmental changes rapidly. This means that other organisms need to adapt more quickly than ever before if they are to avoid **extinction**.

When people began to hunt in groups and use weapons, they became very effective killers. It is possible that the efficient hunting of early people helped wipe out species such as the woolly mammoth and saber-toothed tiger, which could not adapt to cope with the attack.

Biston betularia and the Industrial Revolution

The peppered moth, *Biston betularia*, is found naturally in two different color forms. One is a light, speckled color, and the other is black. Until the 19th century, the pale form was the most common in Great Britain. It was almost invisible on tree bark to the birds that wanted to eat it, so it was much more successful than its darker relative. However, after the Industrial Revolution in Europe, smoke pollution from factories darkened the trees, particularly in the north, and the number of black moths increased rapidly. These dark moths on the dark tree trunks were better camouflaged from the predatory birds. In more recent years, pollution has been better controlled and the tree trunks are pale again. As a result, there are once again more light-colored peppered moths than dark ones.

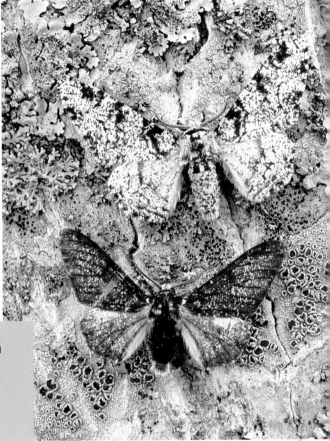

We can see which of the two forms of peppered moth is better camouflaged on a clean tree trunk.
A change in the environment due to human activity resulted in a change in the most common form of peppered moth.

Helping out the frogs

We hear a lot about human destruction of habitats and its effect on different types of animals and plants, but sometimes people do just the opposite and help a species to survive. In the tropical rain forests of Puerto Rico, for example, there is a species of frog that feeds on insects at night and hides from **predators** during the day. Their food is plentiful, so scientists were puzzled when they found that the numbers of frogs were low. Small bamboo shelters were built so that the frogs had more hiding places. As a result, the frogs were able to breed more successfully and their numbers rapidly increased. Bird boxes and bat boxes all over the world are used to do the same thing—increase the breeding success of different animal species.

Good or Bad?

We are constantly affecting the world around us. Is our influence good or bad? It is easy to show how people are damaging the **environment**, but there are many examples of humans in positive roles as well.

The human managers—good . . .

Sand dunes are vulnerable environments in coastal regions around the world. They are damaged too easily and blow away. Conservation work using fencing and planting marram grass can stabilize the dunes and prevent their destruction. Many other positive actions are taken by people all over the world, too. These include setting up national parks and conservation areas, providing nesting boxes, and managing woodlands so that new trees grow before the old ones die.

All over the world there are ecosystems and species of animals and plants that owe their continued existence to the support of human communities. At this tree nursery in Ghana, local workers are looking at young seedlings as part of a forestry project.

. . . or bad?

The production of food is vital. However, the pesticides and herbicides used to help grow crops can be devastating to wildlife if they build up in **food chains**. Also, the fertilizers used to increase crop yields can cause problems in rivers, ponds, and lakes, and sometimes kill the animals that live there.

Many countries in the **developing world**, such as Brazil, are cutting down rain forests to provide wood for developed countries, such as the United States and United Kingdom, to make furniture, ornaments, and even toilet seats. The cleared land is then often used to farm cattle to provide beef for the **developed world**

ost people in these developing untries are desperately poor. They e exploiting the only resource they ve to get food, medicine, and ucation for their families. Yet the fect of rain forest destruction the whole planet is potentially devastating. With the problems of **acid rain** and the build-up of greenhouse gases from our car and factory fumes (the **greenhouse effect**), this shows how serious the effects of some human activity can be on the earth's ecosystems.

Unknown to anyone, waste from a plastics factory in Minamata, Japan, contained high levels of mercury. It was pumped out into the bay and built up in the mud at the bottom of the sea.

B. In the bay, levels of mercury built up in shellfish and fish, but this did not kill them.

C. Local fisherman fed both their families and cats on the fish and shellfish they caught. The cats became ill first, gradually became paralyzed, and finally died. No one could understand why.

D. Finally, people showed the same symptoms. More than 60 people were permanently disabled, 44 people died, and babies were born deformed before the source of the mercury poisoning at the beginning of the food chain was found and stopped.

▲ The Minamata tragedy. In the 1950s, the effects of human pollution traveled along a food chain and devastated the lives of people in the area.

iological pest control

hemical pesticides have caused a lot environmental damage in the last ndred years. On the other hand, ithout the pesticides, many crops ould have failed and millions more eople would have died of starvation. eople are increasingly looking for ways control pests without using toxic hemicals. Biological pest control seems offer one good solution. This involves using an animal (often an insect) or a plant to control pest organisms. For example, you can find ladybugs in your garden and put them on a plant infested with greenfly that the ladybugs will eat. Internationally, research is being done to find suitable, safe organisms to control pests that damage crops or destroy natural environments on a large scale.

Populations—Plants and Animals

Ecology mainly looks at the factors that affect the distribution and abundance of living organisms. This involves studying the success or failure of **populations** rather than single animals or plants. Large numbers give us a much more accurate picture of the way things work.

SCIENCE ESSENTIALS

A population is a group of organisms of the same species living in the same **habitat** at the same time.
The number of organisms in a population will change as a result of factors such as temperature, light, and weather conditions.

What is a population?

A population is a group of similar organisms living in a particular habitat at the same time, producing and consuming food, using up nest sites, and generally making use of the surroundings. In theory, any population will double in a given period of time. If conditions are ideal, it will continue doubling in size to show **exponential growth**. Fortunately, for the balance of life on Earth, ideal conditions rarely exist, and populations grow at fairly slow rates!

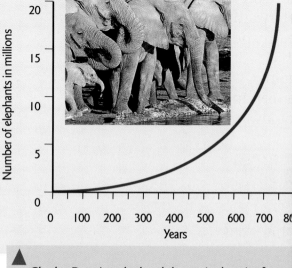

Charles Darwin calculated that a single pair of elephants breeding in ideal conditions would produce 19 million descendants in just 750 years.

Ups and downs of life

The number of birds in a garden can vary greatly from one year to another. The number of coyotes in one area of wilderness can be very different from that in another area only a few miles away. Basic factors such as the type of rocks and soil, the weather, the temperature, and the amount of water and light available all have an effect on population numbers. The weather has a general effect. For example, populations of cacti are generally found where the weather is hot and dry, but tulips grow in temperate climates. Weather can also have very local and severe effects. For example, hurricanes and tornadoes all over the world not only devastate human populations, they also bring death and destruction to tree populations and the animals living in and on them. But in the years that follow, the numbers of fungi, wood-eating insects, and light-loving plants increase sharply and fill the gaps left by the trees.

competition has a big effect on population numbers. It may occur between different members of the same species or members of different species. The interactions between **predators** and **prey** also affect population size. In theory, this is simple. The prey population feeds, grows, and provides food for the predators who, in turn, feed, reproduce, and increase in number. The increased number of predators eat more prey, so the prey numbers fall. Afterward, predators begin to starve to death and the predator numbers fall, which allows prey numbers to rise again. Cycles of numbers like this can be seen in the world. However, it is not always as simple as it seems.

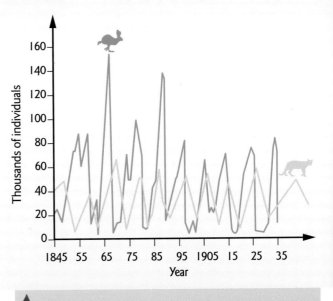

At first, it appears as if the population "peak and crash" cycles of the snowshoe hare and lynx are closely linked together in a classic predator-prey cycle. In fact, the snowshoe hare population rises and falls in the same way in areas where there are no lynxes! The changes are due to fluctuations in the weather and the numbers of plants that the hares need for food.

Population patterns

Some populations of animals and plants have a very high density (a lot of the organisms live very close to each other).

Others have a very low density (only a few organisms live in a big area).

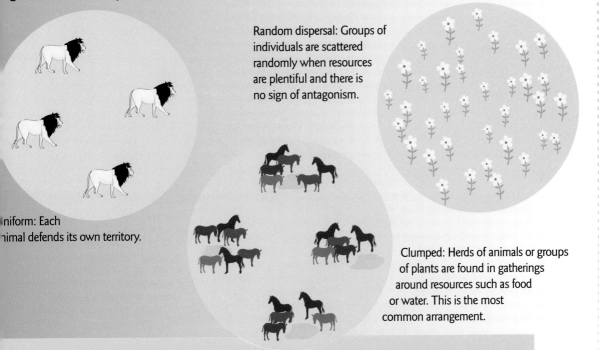

Random dispersal: Groups of individuals are scattered randomly when resources are plentiful and there is no sign of antagonism.

Uniform: Each animal defends its own territory.

Clumped: Herds of animals or groups of plants are found in gatherings around resources such as food or water. This is the most common arrangement.

Depending on both the population density and type of organism, populations of animals and plants are usually spread out in one of these three patterns.

The Human Population

Throughout the history of life on Earth, **populations** have grown and declined. However, for the first time, we now have a population that simply keeps growing.

A human problem

Human beings have lived on Earth for only a short period in the history of our planet. For thousands of years, the human population grew very slowly. Later, improvements in weapons for hunting, tool-making skills, and crop-growing led to increases in the population. This growth was largely canceled out by deaths due to catastrophes, diseases, and wars. At the time of the agricultural revolution in the Middle Ages, it is estimated that there were only about 300 million people in the world. In recent centuries, our ability to grow food and treat diseases has developed rapidly. As a result, the human population has grown enormously. It now has to be measured in billions.

If any other species on Earth experienced a population explosion of the size seen in humans, scientists would predict that an equally dramatic crash would follow. Is that our future too?

Number of people in billions

Years ago in thousands

A force to be reckoned with

The vast human population has spread across the surface of the earth to take advantage of almost every land **habitat**. To maintain our ever-increasing population, we will need to use all of the earth's resources very carefully.

The biggest problem we face is in the production of food. It is not simply a case of growing enough—we do that already. Difficulties lie in producing food where it is needed and in distributing it around the globe. Man people in the **developing world** are largely vegetarian. If the **developed world** ate less meat, there would be a lot more food to go around because the amount of plant food needed to fatten one animal for meat would keep quite a few people from starving

nother problem is the growth of normous cities, particularly in the eveloping world. It is estimated that oon 50 percent of the world's opulation will live in cities. verpopulated cities cause hygiene nd sewage management problems, oor health, and disease. As people ke up more and more space, habitats ill be lost and other animal species ill be pushed out.

ollution of land, air, and water is ound to increase as the human opulation grows. Our only real hope that population growth can be owed down.

▶ More people means that more energy will be used. **Fossil fuels** such as the gasoline we use to run our cars will not last forever. Also, as these fuels are burned, more greenhouse gases are produced. This adds to the **greenhouse effect**.

ringing down the birth rate

many developed countries, the birth te has fallen dramatically over recent ears. Birth control has become easily vailable and women have become ducated and free to lead independent ves. Food and medicine have improved, relatively few children die. Many omen have chosen to have fewer hildren. In many countries, the birth te has dropped below two children er couple, so the population is falling. owever, each child in the developed orld uses far more resources than a hild born in a developing country.

Often in many developing countries, girls are not educated. There may be ignorance about birth control, or it may be forbidden by religious groups. Also, because many children still die from disease and malnutrition, large families are seen as a kind of insurance. And, in many societies, a large family is a status symbol.

It will take many years before people in the developing world feel able to change their views. Meanwhile, the human population climbs ever higher.

Introducing Variety

Millions of species of animals, plants, and microorganisms are alive today. Although they are largely built up of similar cells, they demonstrate the wide variety of organisms on Earth.

Natural selection

People look very different from each other, but they also look different from any other species—so do the members of any other species. The differences between members of the same species are known as **variation**.

When conditions are good, most plant and animal members of a **habitat** will survive. If conditions get tougher, however, it becomes a fight for survival. The most likely to survive have, by chance, a feature that helps them overcome the difficult conditions. Having survived the difficult times, they are then able to reproduce successfully when times improve. Their useful survival trait is passed on to their offspring. This **natural selection** of organisms with the most useful survival traits was called "survival of the **fittest**" by Charles Darwin.

Artificial selection

Not all variety comes about by random natural selection. For thousands of years, people have used selective breeding to develop animals and plants with the features they want. We have made animals tamer, fiercer, smaller, and larger to suit our needs. This involves choosing the parents carefully for the desired features, and then selecting the best of the offspring for further breeding until those features are secured.

▶ Artificial selection has given us cereal crops from flowering grasses and beef and milk cattle from the original wild steers.

All of the vegetables in this picture have been selected for particular features to fulfill a specific need. Internally, they are very similar, but they all look very different.

Nature versus nurture

Variation between organisms is caused partly by the information they inherit from their parents and partly by their **environment**. It is common sense that the surroundings of a living organism will influence the way it turns out, and scientific evidence backs this up.

Most living organisms vary because they have different genetic material. Sometimes clones appear—individuals that are the result of natural or manipulated **asexual reproduction**. Clones are genetically identical. This should mean that any differences between them are the result of differences in their environment. Plants will grow differently if they are deprived of light or water, and animals, including people, will be affected by their diet and other aspects of their environment.

SCIENCE ESSENTIALS

There are differences or "variations" between members of the same species. Because of these variations, some individuals are more successful than others. In harsh or changed conditions, the fittest survive and may eventually form a new species.
Variation between organisms of the same species has **environmental** and **inherited** causes.

Identical twins develop from the same fertilized egg, so they have exactly the same inherited information. However, they don't always look the same. Even identical twins brought up together have a different environment from birth onward. They are born at slightly different times and will always be treated slightly differently. Some twins emphasize the similarities between them by wearing the same clothes. Others like to make sure people notice the differences.

"Pharming" in the Future

Until recently, farming has involved growing plants and animals to provide for human needs: food, drinks, and clothing. But techniques such as **genetic engineering** and **cloning** are changing the way we think about farming.

Genetic engineering

By changing the **genetic** information in an organism, scientists can affect the way the cells and organs work. This has great potential for farmers all over the world. Putting an extra growth gene into animals and plants can make them grow faster and allows food to be produced more cheaply. Genes have been put into tomatoes to keep them from becoming squishy on the supermarket shelves. Genes have been engineered into a variety of crop plants that enable them to produce a natural pesticide. This means that chemical pesticides and their related pollution risks can be avoided. Genetically modified cotton that is exactly the right color for blue denim jeans is being grown. This saves the company the expense of dying the material and avoids the pollution caused by the dying process.

Along with research into new genetically modified products, scientists are also working to make sure they are safe—both for consumers and the **environment**.

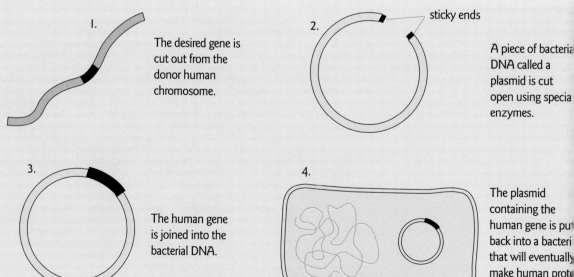

1. The desired gene is cut out from the donor human chromosome.

2. A piece of bacterial DNA called a plasmid is cut open using special enzymes. sticky ends

3. The human gene is joined into the bacterial DNA.

4. The plasmid containing the human gene is put back into a bacteria that will eventually make human protein.

Genetic engineering involves taking a small fragment of DNA (the genetic material) from one organism and "stitching" it into the DNA of another organism.

The "pharm" of the future

Instead of simply farming animals and plants for food and clothing, we are now beginning to "pharm" them—grow them for the pharmacological substances (drugs) that they can produce after genetic engineering. After the addition of a human gene, **bacteria** that produce human insulin for use by diabetics can be grown. Human genes have also been engineered into mammals such as sheep and cattle. The animals then secrete substances in their milk, such as human growth hormone or blood-clotting factors, for treating serious diseases.

Cloning and beyond

Cloning plants by taking cuttings is a long-established practice, but new cloning skills are now being used in several ways. A plant can be broken down into tiny cell fragments and each fragment grown, making thousands of identical clones. Similar techniques are being researched in animals. If animals that have been engineered to make human proteins can be cloned in this way, the benefits spread quickly.

In 1997, the first cloning of a large mammal, Dolly the sheep, was announced. Dolly has gone on to show how normal she is by producing a lamb of her own.

How far, how fast?

Some people accept these new techniques readily because of the many benefits they bring. Others are more cautious. Most genetic engineering involves marker genes that allow scientists to see if the new genes are in place. These markers are often genes for antibiotic resistance, which means an organism given them will not be affected by antibiotics. Nobody is completely sure that this resistance could not be passed into bacteria, giving rise to antibiotic-resistant superbugs capable of causing untreatable diseases. If plants contain pesticide all the time, pests might build up resistance to the poison. There may be problems because people are eating the pesticides along with the plants too. Companies are **patenting** genes and organisms. But is it right for a form of living organism to belong exclusively to a company? Society needs to consider these and other questions sooner rather than later to make sure that the new technologies are a safe approach to the future.

Extinction

Of the estimated 4 billion species that have lived on Earth during its history, only about 2 million are alive today.

Extinction is forever

The great red elk, the Tasmanian tiger, *Tyrannosaurus rex*—today, all of these animals are extinct, even though their names may be familiar.

▲ A landscape like this comes from the mind of the artist. No human has ever seen these dinosaurs alive.

Dinosaurs disappeared from the earth long before our early ancestors even existed.

Why do species of animals or plants become extinct? It often seems to be the result of a change in the **environment**. Many species cope with the change, but others are not so well adapted. Less able competitors do not get as much food or other resources. As a result, they are less likely to reproduce successfully. As their numbers fall, they will be increasingly pushed out by other, better-adapted species. Eventually, the species will become extinct in that area. If the **environmental** change occurs over the whole range of that species, then the whole species may become extinct.

ass extinctions

sides the normal background rate of tinctions, **fossils** suggest that there ve been five **mass extinctions** in e history of the earth. In one of ese, 96 percent of all the marine **vertebrates** died out. In another, up 75 percent of all known plant and imal marine species died out. In the st-known mass extinction, the ighty dinosaurs fell from being dominant animals to being extinct in around 8.5 million years. The effects of these mass extinctions on the evolution of different species should not be underestimated.

After a mass extinction, there are large numbers of empty **habitats**, and new animal and plant species rapidly evolve to fill the empty spaces.

he human effect

a new type of plant or animal evolves, can make other organisms extinct by mpeting more successfully for the me resources. Human beings have iven more species to extinction than y other species has. Since people first olved into social groups, hunting gether and using tools, we have wiped any other species off of the earth.

nopping down rain forests and draining arshes and bogs have caused thousands species to become extinct all over the orld. Polluting river and ocean waters, well as the air around us, has driven any more species to extinction. eople are having an enormous effect the whole environment and the ospects for the future can seem oomy. However, as we become more vare of the problems we are causing, e can look for ways to reduce or event the damage. With people all ver the world working together, the ances of saving the earth for future nerations are greatly improved.

▼

It is thought that early people caused the extinction of many of the species they hunted, including the woolly mammoth. Modern people have continued in the same way. The dodo is one of the best-known species to be driven to exctinction by humans. When sailors first landed on the island of Mauritius, the flightless dodo was unafraid. Unable to escape, it was hunted to extinction for its meat. Hunting, habitat destruction, and pollution continue to cause extinctions around the world every day.

Glossary

abiotic describes the nonliving features of an environment

acid rain rain that is acidic due to dissolved gases such as sulfur dioxide produced by the burning of fossil fuels

adaptation special feature of an organism that enables it to survive in a particular habitat. It may involve the whole organism, parts of the organism, or its behavior.

asexual reproduction way of producing offspring that does not require the joining of two sex cells

bacteria tiny single-celled life forms that do not appear to have a nucleus

biomass amount of biological material

carnivore animal that eats animals

cellulose complex carbohydrate found in plant cell walls

chlorophyll green coloring used by plants to capture the sun's energy

chloroplasts organelles (tiny compartments) in plant cells that contain chlorophyll

circadian rhythm pattern of an organism's life processes that roughly follows a 24-hour cycle

cloning producing identical offspring from the tissue of one organism

competition when organisms compete with each other for the same resources

consumer living organism that needs to eat other organisms to get energy

decomposer specialized consumer, often fungi or bacteria, that breaks down the tissues of a dead plant or animal, and returns the nutrients to the soil

developed world those countries like the United States and much of Europe that have exploited resources in order to gain a relatively high standard of living for their populations

developing world those countries, especially in much of Africa and South America, that are currently preparing their resources for greater exploitation in order to improve the populations' standard of living

DNA (deoxyribonucleic acid) chemical that carries genetic information

dormant when a plant or seed slows its metabolism down to survive difficult conditions

ecology the study of ecosystems

ecosystem all the animals and plants living in an area and the things that affect them, such as the soil and weather

environment an organism's home and its surroundings

environmental linked to the surroundings

enzyme protein molecule in living things that speeds up or slows down reactions in natural chemical processes

evolved changed slowly over long periods of time as adaptations are passed on by the fittest to new generations through breeding

exponential growth when something rapidly increases in size by doubling in a given time

extinction permanent loss of a species of living organism from the earth

feces solid waste passed out of the body

fittest the most well-adapted organism

food chain links between different organisms that feed on each other

food web model of a habitat showing how the animals and plants are interconnected through their feeding habits

fossil remains of a once-living organism preserved in rock, amber, or some other medium

ssil fuel   fuel that has formed over millions of years from the remains of living things

netic   involving the genes (units of inheritance) in the nucleus of a cell

netic engineering   changing the genetic material of an organism

eenhouse effect   warming of the earth's surface as a result of a build-up of greenhouse gases such as carbon dioxide

bitat   place where an animal or plant lives—its home

rbivore   animal that eats plants

bernate   when animals build up body fat during the summer, then go into a very deep sleep during the winter, and keep their body temperature very low

herited   passed on from parents in their genetic material

vertebrate   animal without a backbone

ass extinction   period of time when the great majority of living species become extinct

etabolism   building-up and breaking-down reactions in the biochemistry of a cell or living organism

icrohabitat   very small, localized habitat within a much larger habitat

natural selection   survival of the fittest organisms and the passing on of their genes through reproduction

patenting   registering an idea or product as belonging to a particular individual or organization

photosynthesis   process by which green plants make food from carbon dioxide and water using the sun's energy

population   group of organisms of the same species living in the same habitat at the same time

predator   animal that preys on other animals to obtain food

prey   animal that is hunted and eaten by predators

primary consumers   animals that only eat plants

producer   organism that produces new food by photosynthesis

pyramid of biomass   graphic that shows the amount of biological material in organisms at each level of a food chain

pyramid of numbers   graphic that shows the number of organisms at each level of a food chain

secondary consumer   animal that eats other animals

variation   differences between members of the same species

More Books to Read

ines, John. *Humans & the Environment*. Austin, Tex.: Raintree Steck-Vaughn, 992.

lverstein, Alvin, and Virginia B. Silverstein. *Food Chains*. Brookfield, Conn.: illbrook Press, Inc., 1998.

sar, Jenny. *Endangered Habitats*. New York: Facts on File, 1992.

alker, Jane. *Vanishing Habitats & Species*. Danbury, Conn.: Franklin Watts, 1993.

Index